AGRIBUSINESS MANAGEMENT AND ENTREPRENEURSHIP

SMILE WELLBECK

GRATITUDE

I hope this note finds you well! I wanted to take a moment to extend my heartfelt thanks for choosing to explore "Agribusiness Management and Entrepreneurship." Your decision to delve into the world of agribusiness is not just appreciated but also incredibly inspiring.

In this book, I've poured my passion for agriculture and business into every chapter, aiming to make the complexities of agribusiness management accessible and actionable for you. Whether you're a seasoned entrepreneur or just starting out in the field, I've strived to provide practical insights, real-life examples, and strategies that can empower you to navigate and succeed in this dynamic industry.

Your interest in this subject not only supports my work but also encourages me to continue sharing knowledge and experiences that can make a positive impact in the agricultural community. I truly hope you find valuable information and ideas that resonate with your goals and aspirations.

Once again, thank you for choosing "Agribusiness Management and Entrepreneurship." Your engagement means the world to me. Should you have any thoughts, feedback, or questions as you read through the book, please don't hesitate to reach out. I'm here to support you on your journey in agribusiness.

Copyright © 2024, Smile Wellbeck.

This work and its content are protected under international copyright laws.

No part of this publication may be reproduced, distributed, or transmitted in any form or by any means, including photocopying, recording, or other electronic or mechanical methods, without the prior written permission of the author, except in the case of brief quotations embodied in critical reviews and certain other noncommercial uses permitted by copyright law.

TABLE OF CONTENTS

ABOUT THIS GUIDE

1. INTRODUCTION TO AGRIBUSINESS MANAGEMENT AND ENTREPRENEURSHIP
- DEFINING AGRIBUSINESS MANAGEMENT
- UNDERSTANDING ENTREPRENEURSHIP IN AGRIBUSINESS

2. AGRIBUSINESS MANAGEMENT
- STRATEGIC PLANNING
- ACCOUNTING AND FINANCIAL MANAGEMENT
- MARKETING MANAGEMENT AND DISTRIBUTION
- HUMAN RESOURCE MANAGEMENT
- PRODUCTION MANAGEMENT
- RISK AND CRISIS MANAGEMENT

3. AGRICULTURAL ENTREPRENEURSHIP
- IDEATION, INNOVATION, AND CREATIVITY

- MARKET RESEARCH AND ANALYSIS
- BUSINESS PLANNING
- LEGAL AND REGULATORY FRAMEWORKS
- FUNDING AND FINANCING
- SCALING UP

4. AGRIBUSINESS AND ECONOMIC DEVELOPMENT
- ROLE OF AGRIBUSINESS IN SUSTAINABLE DEVELOPMENT
- AGRIBUSINESS DEVELOPMENT STRATEGIES
- PARTNERSHIP AND COLLABORATION FOR AGRIBUSINESS DEVELOPMENT
- MANAGING EXTERNAL FACTORS AFFECTING AGRIBUSINESS

5. TECHNOLOGY AND INNOVATION IN AGRIBUSINESS
- OVERVIEW OF AGTECH AND DIGITAL AGRICULTURE
- APPLICATION OF TECHNOLOGY IN AGRIBUSINESS MANAGEMENT

- AGTECH STARTUPS AND THE ENTREPRENEURSHIP ECOSYSTEM

6. CASE STUDIES IN AGRIBUSINESS MANAGEMENT AND ENTREPRENEURSHIP
- SUCCESSFUL AGRIBUSINESS MANAGEMENT AND ENTREPRENEURSHIP IN DEVELOPED COUNTRIES
- BEST PRACTICES AND LESSONS LEARNED FROM DEVELOPING COUNTRIES

7. CONCLUSION AND FUTURE TRENDS IN AGRIBUSINESS MANAGEMENT AND ENTREPRENEURSHIP
- SUMMARY OF KEY LEARNING POINTS
- TRENDS AND INNOVATIONS IN AGRIBUSINESS
- RECOMMENDATIONS FOR POLICY MAKERS AND STAKEHOLDERS.

ABOUT THIS GUIDE

Welcome! This guide is your companion through the exciting world of agribusiness management and entrepreneurship. Whether you're a budding entrepreneur or someone deeply passionate about agriculture, this book is tailored to help you navigate the complexities of running and growing a successful agribusiness.

Here, you'll find practical insights and strategies that real farmers and business owners use every day. From understanding market trends to mastering financial management in agriculture, each chapter is designed to equip you with the knowledge and skills needed to thrive in this dynamic field.

Think of this guide as your trusted advisor, offering not just theoretical knowledge but also actionable tips and case studies from the field. Whether you're interested in starting your own farm-to-table venture or optimizing existing agricultural operations, you'll find valuable guidance here.

So, dive in and explore how agribusiness management and entrepreneurship can transform your passion for agriculture into a sustainable and profitable enterprise. Let's grow together!

1. INTRODUCTION TO AGRIBUSINESS MANAGEMENT AND ENTREPRENEURSHIP

DEFINING AGRIBUSINESS MANAGEMENT

Agribusiness management is an integral component of the agricultural industry. It is the scientific approach to managing an agribusiness enterprise that entails the coordination of resources and activities involved in the production, processing, and marketing of agricultural products.

In practical terms, agribusiness management combines agriculture, technology, and business management principles to maximize the efficiency of an agricultural enterprise. Successful agribusiness managers utilize various techniques to optimize production, enhance profits, and mitigate risks, all while ensuring the sustainability of the business and the environment.

In today's world, where agricultural practices are being modernized with technology, agribusiness management has become an essential part of the industry. It helps agribusiness owners to become better managers of their land and farms by providing them with tools and knowledge to implement successful farming techniques.

One of the real-life situations in which agribusiness management can be applied is the management of a crop plantation. For instance, let's take a look at a farmer who manages a maize plantation. Through the implementation of agribusiness management techniques, the farmer can optimize various processes such as soil preparation, crop rotation, fertilization, irrigation, and harvesting to maximize yields, mitigate risks, and boost profits.

The farmer also needs to understand the market trends and adjust the production schedule accordingly. By constantly analyzing the prices of maize in local and international markets, the farmer can forecast demand and supply, and make informed decisions on when to plant, when to harvest, and how to market the crop.

Moreover, the agribusiness manager must always be on the cutting edge of innovation. By being aware of the latest technology and techniques, the manager can optimize processes and reduce production costs, which ultimately translates into increased profits.

Agribusiness management is a vital component of the agricultural industry. By providing farmers and agribusiness owners with tools and techniques to optimize processes, manage risks, and enhance profits, it contributes to the sustainability and profitability of agricultural enterprises. With the world's population increasing rapidly, it is crucial that agribusiness embraces best practices and modern techniques to meet the growing demand for food production. Let's all embrace these ideas to not only become better farmers but to also contribute positively to the society and by extension the world.

UNDERSTANDING ENTREPRENEURSHIP IN AGRIBUSINESS

As the world's population continues to grow, it has become increasingly important to find sustainable ways to feed everyone. That's where agribusiness comes in - the practice of managing agriculture and food production and distribution. But what is entrepreneurship in agribusiness?

Entrepreneurship in agribusiness involves developing innovative ideas and approaches to enhance and improve agricultural practices. It's about taking risks and seizing opportunities to create value in the industry, whether it be through introducing new technologies or marketing strategies.

One of the best examples of entrepreneurship in agribusiness is the rise of urban agriculture. With cities becoming more crowded and space becoming more scarce, urban farmers have found innovative ways to grow food in small spaces such as unused rooftops and balconies. This is a prime example of how entrepreneurs are altering traditional methods of food production and distribution.

Another example is the use of precision agriculture technologies such as drones and sensors to optimize crop yields and reduce environmental impact. These advancements not only benefit farmers, but also lead to more efficient use of resources and increased production that can feed more people.

But entrepreneurship in agribusiness is not just about groundbreaking innovations. It also involves finding ways to improve existing practices and make them more sustainable. For example, rather than using harmful pesticides and fertilizers, farmers can adopt regenerative methods that promote soil health and biodiversity while producing healthy and nutritious crops.

Being an entrepreneur in agribusiness means being adaptable, creative, and willing to take risks. It's about finding solutions to complex challenges and contributing to a more sustainable future.

Understanding entrepreneurship in agribusiness is critical in developing innovative and sustainable solutions to feed the growing population. By embracing new technologies and creative marketing approaches, farmers can create value and stay competitive in the industry while minimizing environmental impact. It is the time to explore and welcome these innovative ways of farming into our lives.

2. AGRIBUSINESS MANAGEMENT

STRATEGIC PLANNING

Welcome to the interesting world of strategic planning! As the name implies, strategic planning involves creating an action plan that is designed to achieve a long-term goal in a structured and systematic way. It involves taking a systematic approach to analyzing both external and internal variables that affect the accomplishment of your goals. This is important in agribusiness management as it helps in identifying opportunities and potential challenges faced within the agricultural industry.

In the world of agribusiness management, a company that creates and implements a well-crafted strategic plan is better positioned to achieve success. Strategic planning ultimately determines the direction and scope of your business in the long run. It will guide your decision-making when it comes to what to do and what not to do.

Creating a strategic plan involves taking a comprehensive approach, which entails identifying your business's strengths, weaknesses, opportunities, and threats. You also need to take a close look at the market conditions, identify emerging trends, and predict future patterns of demand and supply within the agricultural space.

When creating a strategic plan, it is important to have a clear understanding of your company's mission and vision, and how that aligns with the overall goals of the agricultural industry. A well-crafted plan should be based on accurate data and research, and should be flexible enough to change with the changing needs of your business and the market.

In order to create an effective strategic plan in agribusiness management, good communication is key. It is important that all stakeholders involved in the strategic planning process are involved and feel valued. This ensures that the whole team is aligned and working towards the same vision, with a clear understanding of the goals and priorities at hand.

Strategic planning is an essential tool for success in agribusiness management. By taking a comprehensive and systematic approach, you can identify opportunities and challenges, and develop an action plan to achieve your long-term goals. With collaboration and communication, the whole team can come together to execute a well-crafted strategic plan successfully. By investing time and resources in strategic planning, you can take your business to the next level and achieve long-term success in the agricultural industry.

ACCOUNTING AND FINANCIAL MANAGEMENT

To pursue informed choices and drive development, fundamental to take on sound monetary systems will assist you with exploring the unique idea of the farming business. Accounting and financial management are a portion of the key support points that will guarantee your agribusiness runs productively and really.

Bookkeeping fills in as the groundwork of Financial management, offering a preview of your business' ongoing monetary position. It assists with following monetary exchanges, keeping them in a way that precisely mirrors your business' monetary position. By keeping appropriate monetary records, you can evaluate the benefit of your agribusiness, go with very much educated choices, and gain bits of knowledge into regions that require improvement.

Financial management, then again, includes utilizing monetary information to go with key choices that benefit the business. It centers around the monetary prosperity of your agribusiness, from gauging and planning to monetary examination and control. Sound Financial management permits you to decide the feasibility of a venture, open doors, oversee monetary dangers and recognize areas of shortcoming that can be redressed.

With regards to agribusiness management, Accounting and financial management rehearses are fundamental in the everyday running of your venture. Without legitimate Financial management, your business could run into critical issues that might actually harm your standing, dissolve your overall revenues and even lead to the conclusion of the business. By embracing powerful Accounting and financial management rehearses, you'll have the option to settle on informed choices, plan for development, and keep your business reasonable in a dynamic and testing industry.

Genuine uses of Accounting and financial management rehearses in agribusiness management are various. For instance, you could utilize bookkeeping reports to follow costs, screen deals and benefits, and make changes on a case by case basis. You could likewise utilize Financial management apparatuses to oversee income, recognize speculation, open doors, and foster spending plans and monetary projections.

Accounting and financial management are basic parts of agribusiness. In the present dynamic and always changing agrarian industry, pivotal to embrace powerful monetary systems will assist you with overseeing gambles, examine market patterns, and settle on informed choices. By improving your Accounting and financial management abilities, you'll carry more worth to your business, position it for development, and guarantee its supportability as long as possible.

MARKETING MANAGEMENT AND DISTRIBUTION

Showcasing management and appropriation are fundamental parts of agribusiness the board. To prevail in the business, it is fundamental to actually advance horticultural items and convey them effectively to target markets.

Marketing management includes creating systems to encourage an interest for farming items while recognizing them from contenders. It includes leading statistical surveying to comprehend shopper necessities, inclinations, and patterns. By understanding what shoppers are searching for, agribusinesses can fit their showcasing procedures to address those issues.

One illustration of successful showcasing in agribusiness is the advancement of natural items. Numerous customers are worried about the wellbeing and ecological effects of customary cultivating techniques. By utilizing this worry and stressing the advantages of natural cultivating, organizations can draw in purchasers who focus on manageability and wellbeing.

Distribution is one more significant part of agribusiness, management. It includes deciding the most proficient and savvy strategies for getting items from the maker to the purchaser. The topographical area of the maker and target market are key factors that decide the method of dissemination.

One illustration of proficient conveyance in agribusiness is the utilization of rancher's business sectors. These business sectors give an immediate connection among ranchers and shoppers, taking into consideration new, privately created harvests to be sold straightforwardly to buyers. This additionally assists with decreasing the natural effect of transportation and capacity.

Notwithstanding conventional promoting and distribution techniques, there is a developing pattern towards online business in agribusiness. This permits makers to contact a more extensive crowd and opens up new showcasing and circulation channels. A perfect representation of web based business in agribusiness is the internet requesting and conveyance of new produce straightforwardly to purchasers' homes.

Marketing management and distribution are basic parts of agribusiness management. Viable promoting methodologies and proficient conveyance techniques can assist agribusinesses with hanging out in a packed industry and arrive at new clients. By understanding purchaser needs and utilizing innovation, agribusinesses can work on their main concern while advancing supportable and solid horticultural practices.

HUMAN RESOURCE MANAGEMENT

Human resource management is a basic part of Agribusiness management. Agribusiness is an industry that includes the creation, handling, and circulation of farming items. To prevail in this industry, it is essential to have gifted and committed representatives who can cooperate really to accomplish the organization's targets. This is where human asset the board comes in.

Viable human resource, management includes selecting and recruiting the ideal individuals to get everything taken care of, giving them legitimate preparation and improvement, and establishing a rousing workplace that advances development and worker maintenance. In the Agribusiness business, it is fundamental to have representatives who have the necessary abilities and experience as well as figure out the idea of the business.

Enrolling and holding workers in the Agribusiness business can be a test; nonetheless, there are systems that organizations can use to address this. Offering cutthroat wages and advantages is one of the best ways of drawing in and holding workers. Workers are likewise bound to remain in an organization on the off chance that they feel their occupation is significant, and their commitments are esteemed.

Preparing and advancement are additionally basic parts of human asset the board in the Agribusiness business. The business is continually advancing, and staying up with the latest with the most recent patterns and technologies is fundamental. This will empower them to play out their obligations all the more really and effectively. Organizations can utilize different preparation procedures, for example, hands-on preparing, study hall preparing, e-learning, or training and tutoring programs.

Establishing a positive workplace is one more fundamental consideration of human assets. Positive workplaces empower worker commitment and advance a feeling of belongingness, bringing about expanded work fulfillment and efficiency. To establish a positive workplace, organizations can support joint effort, representative contribution in dynamic cycles, and continuous worker criticism.

Human resource management is an urgent part of Agribusiness management. It includes enlisting and holding gifted representatives, giving them the important preparation and improvement, and establishing a positive workplace. The Agribusiness business has novel difficulties, yet with the reception of the right human asset management rehearse, organizations can fabricate a gifted, persuaded, and drawn in labor force.

PRODUCTION MANAGEMENT

Production management is a basic part of any agribusiness, as it decides the quality, amount, and cost of farming items. It includes arranging, sorting out, coordinating, and controlling assets to accomplish creation objectives and goals.

Consider a rancher who establishes a harvest, like maize. The rancher should deal with all parts of the creation cycle, from planting to reap. All that from choosing the right seeds, setting up the land, planting, applying manures and pesticides, and the water system should be overseen accurately and proficiently.

To start, the rancher needs to settle on the right seeds to plant. They ought to consider factors, for example, environment, soil type, seed cost, and yield potential. When the right seeds have been chosen, they need to set up the land for planting. This can include clearing weeds, plowing the dirt, and adding composts or naturally making a difference to the dirt.

Subsequent to planting, the rancher needs to deal with the development of the yield by observing for nuisances and illnesses, applying manures and herbicides, and guaranteeing the harvest has the right circumstances for development. This incorporates sufficient water and daylight.

In the long run, the opportunity will come to reap the yield. Appropriate preparation of the father is fundamental to guarantee that the yield is reaped brilliantly, utilizing the right methods, and put away actually to keep away from waste. The rancher needs to choose conventional manual reaping or utilizing machines that might accelerate the interaction.

As well as dealing with the creation interaction, the rancher likewise needs to consider the market interest and plan ahead on the most proficient method to sell and appropriate the produce. This implies distinguishing likely purchasers, guaranteeing they meet the important necessities, and setting the right costs.

Fruitful creation of management is vital for any agribusiness since it decides the quality, amount, and cost of farming items. With successful creation of the board, ranchers can augment yields, limit costs, and guarantee that their items fulfill the vital quality guidelines.

RISK AND CRISIS MANAGEMENT

With regards to agribusiness management, risk and crises management are pivotal parts. In the farming business, dangers like catastrophic events, market vulnerability, and sickness episodes can essentially affect a business' benefit and supportability. Emergency the board, then again, includes overseeing surprising occasions that might actually harm the standing of the business or lead to huge misfortunes.

One vital part of risk and crises management in agribusiness is recognizing likely dangers and creating techniques to alleviate them. For instance, in the event that a business is situated in a space inclined to cataclysmic events, for example, floods, it's critical to do whatever it takes to safeguard the business and its resources. This could include fostering a crisis reaction plan, putting resources into protection inclusion that covers risks like flooding, and doing whatever it may take to sustain structures and different resources against possible harm.

As well as recognizing likely dangers, organizations ought to likewise have an emergency the board plan set up. An emergency could be anything from an item review to a sickness flare-up, and it's fundamental to be ready to answer rapidly and successfully. This incorporates having an unmistakable arrangement for speaking with partners, creating emergency courses of action for basic business works, and guaranteeing that staff individuals are prepared to answer in an emergency.

Genuine uses of risk and crises management in agribusiness management remember the 2001 foot-and-mouth sickness flare-up for the UK. This emergency upset the cultivating business as well as fundamentally affected the travel industry, prompting misfortunes of billions of dollars. The emergency exposed the significance of chance and emergency to management in the agrarian business and prompted the improvement of additional powerful techniques for managing flare-ups from here on out.

Another model is the 2020 Coronavirus pandemic, which altogether affected the horticultural business, including disturbed supply chains and falling interest for specific items. Organizations that had created strong gambling and emergency board plans were better prepared to face the hardship, while those that had not were hit a lot harder.

Risk and crises management are basic parts of agribusiness the board. Recognizing likely dangers, creating methodologies to alleviate them, and having compelling emergencies management' plans set up are fundamental for guaranteeing the maintainability and progress of a farming business.

3.0 AGRICULTURAL ENTREPRENEURSHIP

IDEATION, INNOVATION, AND CREATIVITY

In the vast and ever-evolving landscape of agricultural entrepreneurship, three essential elements stand out as the driving forces behind meaningful progress and sustainable growth: ideation, innovation, and creativity. These elements are not just theoretical concepts but practical tools that empower farmers and agricultural entrepreneurs to navigate challenges, seize opportunities, and make a lasting impact in their communities.

Ideation: At the heart of every successful agricultural venture lies a seed of an idea – a solution to a problem, an improvement to an existing practice, or a completely new approach to farming. Ideation begins with observation and empathy, understanding the needs and aspirations of farmers and consumers alike. For instance, a farmer noticing increased soil erosion due to heavy rains might ideate on implementing contour farming techniques to preserve soil fertility and water retention. Ideation is about being attentive to the nuances of the agricultural environment, drawing inspiration from traditional wisdom, scientific research, and practical experience to innovate solutions that are both effective and sustainable.

Innovation: Once an idea takes root, innovation is the process of refining and implementing it to create real-world impact. Innovation in agricultural entrepreneurship spans a spectrum of activities – from adopting new technologies and practices to adapting existing solutions to local contexts. For example, a small-scale dairy farmer innovating by introducing a mobile milking unit to reduce labor costs and improve hygiene standards. Innovation is not limited to technological advancements but also encompasses improvements in business models, marketing strategies, and supply chain management that enhance efficiency, productivity, and profitability.

Creativity: Creativity is the spark that transforms ideas and innovations into tangible outcomes. It's about thinking outside the box, taking calculated risks, and embracing experimentation. Creative agricultural entrepreneurs are those who find unconventional solutions to common challenges, such as using recycled materials for greenhouse construction or organizing agritourism activities to diversify income streams. Creativity thrives on collaboration and multidisciplinary approaches, drawing inspiration from diverse fields such as ecology, economics, and sociology to address complex agricultural issues.

Consider the story of a farmer in a drought-prone region who, through ideation, developed a rainwater harvesting system to ensure a consistent water supply for irrigation throughout the year. His innovation not only improved crop yields but also inspired neighboring farmers to adopt similar practices, fostering community resilience and sustainability.

Another example could be a group of young entrepreneurs who identified a niche market for organic produce in urban areas and creatively established a cooperative farming model that connects urban consumers directly with rural producers. This initiative not only promotes sustainable farming practices but also strengthens local economies and promotes food security.

Ideation, innovation, and creativity are not just buzzwords but essential pillars of success in agricultural entrepreneurship. They empower individuals and communities to embrace change, adapt to challenges, and unlock new opportunities in the agricultural sector. By fostering a culture of continuous learning, collaboration, and resilience, agricultural entrepreneurs can harness the full potential of ideation, innovation, and creativity to create a future where farming is sustainable, profitable, and enriching for all stakeholders involved.

MARKET RESEARCH AND ANALYSIS

In the realm of agricultural entrepreneurship, success hinges not only on the quality of the produce but also on a deep understanding of market dynamics. Market research and analysis serve as the bedrock upon which sustainable agricultural ventures thrive, guiding entrepreneurs to make informed decisions that resonate with consumer demands and market trends.

Imagine embarking on a journey to establish a farm specializing in organic vegetables. Your passion for sustainability and dedication to providing nutritious, pesticide-free produce drive your ambition. However, without a clear understanding of your target market, the path forward remains obscured.

Understanding Your Consumer Base

Market research empowers agricultural entrepreneurs to identify and understand their consumer base. It involves delving into demographics, preferences, purchasing behaviors, and even cultural nuances that influence buying decisions. For instance, you might discover that urban consumers are increasingly inclined towards locally sourced, organic products due to health and environmental concerns. Armed with this insight, you can tailor your farming practices and marketing strategies accordingly, ensuring alignment with consumer expectations.

Navigating Market Trends

The agricultural landscape is dynamic, shaped by evolving trends and external factors such as climate change, economic fluctuations, and technological advancements. Market analysis enables entrepreneurs to anticipate these shifts and adapt proactively. For instance, an analysis might reveal a growing demand for heirloom tomatoes or an emerging market for farm-to-table experiences. By staying abreast of such trends, you position your venture to capitalize on emerging opportunities and mitigate potential risks.

Identifying Competitive Advantages

In a competitive market, differentiation is key to success. Through rigorous market analysis, entrepreneurs gain valuable insights into their competitors' strengths and weaknesses. This knowledge not only helps in identifying gaps in the market but also enables the development of unique value propositions. For instance, you might discover that neighboring farms lack the infrastructure for direct-to-consumer sales, presenting an opportunity to offer convenient online ordering and home delivery services.

Mitigating Risks and Enhancing Viability

Entrepreneurship inherently involves risks, and agricultural ventures are no exception. Market research serves as a strategic tool for risk mitigation by providing data-driven insights that inform decisions related to pricing strategies, production planning, and investment priorities. For example, thorough analysis might reveal seasonal fluctuations in demand for specific crops, prompting you to diversify your product offerings or implement crop rotation practices to optimize yield and revenue stability.

Building Sustainable Growth

Ultimately, market research and analysis foster sustainable growth by fostering a deep understanding of market dynamics, consumer preferences, and competitive landscapes. This holistic approach not only enhances the viability of agricultural ventures but also cultivates resilience in the face of uncertainty. By continuously refining strategies based on real-time market insights, entrepreneurs can nurture thriving agricultural enterprises that contribute to local economies, promote environmental stewardship, and meet the evolving needs of consumers.

Market research and analysis are indispensable tools for agricultural entrepreneurs aspiring to carve out a niche in a competitive market. By embracing these practices, you embark on a journey of discovery and innovation, paving the way for sustainable success rooted in a deep understanding of market dynamics and consumer behaviors.

BUSINESS PLANNING

In the domain of horticultural business, the excursion from seed to gathering is complicatedly woven with the strings of fastidious preparation and vital prescience. Business planning remains as the foundation that changes yearnings into substantial accomplishments, directing horticultural business visionaries through the intricacies of development, market elements, and reasonable development.

Making Your Vision

At the core of each and every effective farming endeavor lies a convincing vision. Imagine yourself remaining on a plot of land, imagining columns of prospering yields or a clamoring ranch to-table activity. Business planning starts with articulating this vision; a guide that depicts your central goal, values, and long haul yearnings. By characterizing clear targets and adjusting them to your enthusiasm for agribusiness, you set up a reason driven business venture that resonates with partners and shoppers the same.

Delineating Strategies

Horticultural business venture requests a nuanced way to deal with procedure definition, enveloping creation methods, market entrance, and functional efficiencies.

Business planning works with the delineating of these techniques, drawing on experiences gathered from statistical surveying and investigation. For example, you could frame development rehearses improved for neighborhood soil conditions or devise dispersion channels that take special care of metropolitan customers' inclination for new, privately obtained produce. By coordinating supportable practices and utilizing mechanical developments, you upgrade efficiency as well as relieve ecological effect; a demonstration of your obligation to dependable stewardship.

Monetary Reasonability and Asset Allocation

Exploring the monetary scene is crucial in horticultural business, where irregularity and outer variables have an effect on productivity. Business planning ingrains monetary judiciousness by estimating income streams, planning costs, and distinguishing wellsprings of financing or speculation. Consider the situation of getting support for extending nursery framework or putting resources into cutting edge water system frameworks to advance water use. By fastidiously overseeing assets and assessing profit from speculation, you strengthen the monetary versatility of your endeavor, preparing for practical development and functional coherence.

Building Versatility Through Hazard Management
In farming, versatility is developed through proactive gambling the board methodologies implanted inside exhaustive strategies. From climate variances and irritation flare-ups to advertise instability and administrative changes, agrarian business visionaries face a variety of vulnerabilities. Business planning engages you to expect and moderate these

dangers by executing alternate courses of action, differentiating item contributions, and cultivating vital organizations. For example, you could lay out elective stockpile chains or put resources into crop protection to shield against unanticipated difficulties, accordingly sustaining the flexibility of your agrarian undertaking.

Cultivating Partner Commitment and Local area Impact

Horticultural business venture flourishes inside interconnected environments of partners; going from providers and merchants to nearby networks and customers. Business planning encourages partner commitment by developing straightforward correspondence, moral practices, and social obligation. Envision teaming up with nearby schools to instruct youngsters about maintainable cultivating rehearsals or cooperating with local area associations to give overflow produce to food banks. By sustaining significant connections and contributing decidedly to the local area, you fashion a strong organization of help that reinforces the life span and effect of your farming endeavor.

Embracing Advancement and Adaptation
In a period characterized by fast mechanical progressions and developing buyer inclinations, dexterity and advancement are impetuses for agrarian business ventures. Business arranging supports ceaseless development by embracing rising advancements, for example, accuracy in horticulture and information investigation, to upgrade yields and functional efficiencies. Envision coordinating IoT sensors to screen soil dampness levels or taking on blockchain innovation to upgrade discernibility and straightforwardness in the production network. By remaining in front of industry patterns and embracing versatile systems,

you position your farming endeavor at the cutting edge of development, driving manageable development and upper hand.

Business planning rises above simple documentation; it exemplifies the essential premonition, versatility, and responsibility that characterize effective horticultural business ventures. By creating a vigorous strategy established in vision, procedure, monetary reasonability, risk management, partner commitment, and development, you leave on a groundbreaking excursion towards understanding your horticultural yearnings. With each season, you develop crops as well as reasonable achievement that leaves an enduring heritage in the horticultural scene.

LEGAL AND REGULATORY FRAMEWORKS

In the powerful domain of horticultural business venture, exploring the legal and regulatory frameworks isn't simply a question of consistency; it is a basic point of support for maintainable development and development. These structures structure the bedrock whereupon agrarian endeavors can prosper, guaranteeing decency, security, and natural obligation in each part of activity.

Guaranteeing Decency and Value

Lawful systems in horticulture are intended to maintain reasonableness for all partners required, from limited scope ranchers to enormous rural ventures. They lay out rules for fair rivalry, forestalling monopolistic practices that could smother development and cut off market access. By guaranteeing a level battleground, these guidelines encourage a climate where innovative souls can flourish, no matter what the size or size of the endeavor.

Shielding Quality and Security

Quality and security principles are foremost in rural business. Administrative bodies put forward rigid rules for the creation, dealing with, and appropriation of rural items. These norms shield customers from destructive toxins as well as improve the standing of agrarian business visionaries by ensuring the most elevated levels of value in their contributions. Consistence with these guidelines isn't simply a legitimate prerequisite however a guarantee to greatness and customer trust.

Advancing Natural Obligation

In the advanced period, ecological manageability is non-debatable in horticultural practices. Administrative structures order rehearses that limit

52

ecological effect, like water protection, soil wellbeing the board, and biodiversity conservation. By incorporating these practices into ordinary activities, farming business people follow legitimate guidelines as well as add to the drawn out soundness of the planet. This obligation to supportability is progressively esteemed by customers and financial backers the same, situating farming endeavors as dependable stewards of normal assets.

Encouraging Advancement and Transformation

As opposed to normal discernment, legitimate systems in farming are not prohibitive; they are empowering agents of advancement. By giving clear rules and motivators, these guidelines urge business visionaries to investigate new advances, techniques, and plans of action that can reform the business. Whether it's utilizing computerized answers for accuracy cultivating or taking on natural cultivating rehearsals, business visionaries are engaged to spearhead arrangements that fulfill advancing shopper needs and administrative necessities.

Moderating Dangers and Upgrading Versatility

Business intrinsically implies gambles, and agrarian endeavors are no exemption. Legitimate structures assume an urgent part in moderating these dangers by offering securities against unanticipated conditions like market variances, catastrophic events, or legally binding questions. By giving an organized structure to gamble with management and debate goals, these guidelines support the versatility of horticultural business visionaries, empowering them to explore difficulties with certainty and coherence.

Legal and regulatory frameworks are not just regulatory obstacles; they are fundamental foundations of progress in horticultural business. Embracing these structures guarantees consistent with the law as well as lays the preparation for long haul development, benefit, and positive effect on networks and the climate. As rural business ventures keep on developing, the job of legitimate and administrative systems will stay basic, molding a future where development and supportability remain closely connected.

FUNDING AND FINANCING

Getting funding and financing isn't simply an issue of financial exchanges; a basic life saver decides the achievement and supportability of adventures. From seed subsidizing for imaginative new companies to credits for laid out ranches hoping to grow, exploring the scene of monetary help requires key preparation, strength, and a profound comprehension of accessible assets.

Seed Financing: Supporting Advancement From the beginning

For yearning rural business visionaries with historic thoughts, seed subsidizing fills in as the underlying flash that touches off advancement. Whether it's growing new yield assortments, spearheading reasonable cultivating practices, or sending off ranch to-table drives, seed subsidizing gives the capital expected to change thoughts into substantial endeavors. This beginning phase funding frequently comes from private backers, financial speculators, or government awards intended to help advancement in horticulture. Getting seed subsidizing isn't just about the cash; it's tied in with picking up approval and speed to impel the endeavor forward in the midst of the vulnerabilities of beginning phase improvement.

Advances and Credit: Energizing Development and Extension

As agrarian endeavors mature and scale, admittance to advances and credit becomes fundamental for supporting development and immediately taking advantage of chances. Whether it's obtaining new gear, extending creation limits, or putting resources into advertising and conveyance channels, credits give the monetary influence expected to benefit from market interest and amplify efficiency. Farming business people can get advances through banks, credit associations, or specific rural moneylenders that grasp the novel dangers and occasional nature of the business. These monetary instruments give quick liquidity as well as cultivate long haul dependability by empowering key interests in foundation and innovation.

Government Awards and Sponsorships: Encouraging Strength and Maintainability
Legislatures overall perceive the crucial job of horticulture in food security, financial dependability, and natural stewardship. In that capacity, they offer awards, appropriations, and impetuses to help rural business. These monetary assets mean to lighten startup costs, advance supportable cultivating practices, and upgrade flexibility against market instability and environmental change. From sponsorships for natural certificates to awards for sustainable power projects on ranches, government support assumes an urgent part in empowering business people to take on imaginative arrangements and contribute emphatically to the farming environment.

Influence Effective money management: Connecting Benefit with Reason

As of late, effective money management has arisen as a strong power in horticultural business, adjusting monetary re-visitations for social and natural effect. Influence financial backers look for amazing open doors that create positive results past monetary benefit, like further developing occupations in provincial networks, advancing biodiversity protection, or moderating environmental change through regenerative cultivating rehearses. By drawing in influence from financial backers, horticultural business visionaries secure capital as well as get to mastery, organizations, and organizations that enhance their effect and maintainability endeavors.

Crowdfunding and Local area Backing: Outfitting Aggregate Strength

In the computerized age, crowdfunding stages have democratized admittance to capital for agrarian business visionaries. By sharing their accounts and dreams on the web, business visionaries can draw in help from a worldwide local area of benefactors who put stock in their central goal. Crowdfunding gives monetary assets as well as develops a dedicated client base and supporters who champion the endeavor's prosperity. Moreover, people groups upheld horticulture (CSA) models permit shoppers to put straightforwardly in neighborhood ranches, encouraging a harmonious relationship based on trust, straightforwardness, and shared values.

Subsidizing and funding are the soul of horticultural business, controlling development, development, and supportability in a powerful industry. From seed subsidizing for intense plans to credits for development and government support for versatility, the accessibility of different monetary assets engages business visionaries to explore difficulties, jump all over chances, and have a significant effect on worldwide food frameworks. As the scene keeps on developing, the mission for practical financing arrangements stays vital to molding a future where horticulture flourishes as a foundation of monetary thriving and ecological stewardship.

SCALING UP

Scaling up in agricultural entrepreneurship marks a pivotal journey from humble beginnings to substantial impact, where small ventures evolve into robust enterprises capable of meeting broader market demands while maintaining integrity and sustainability.

Vision and Strategic Planning: Charting the Course for Growth

Scaling up begins with a clear vision and strategic planning. It's about envisioning not just where the business stands today, but where it can go tomorrow. For agricultural entrepreneurs, this vision might involve expanding production capabilities, diversifying product lines, or penetrating new markets. Strategic planning involves setting achievable goals, assessing market opportunities, understanding consumer needs, and identifying potential challenges. It's a roadmap that ensures every step forward aligns with the overarching mission and values of the enterprise.

Operational Efficiency: Streamlining Processes for Growth

As agricultural ventures scale, operational efficiency becomes paramount. This entails optimizing workflows, investing in technology and infrastructure, and implementing best practices to maximize productivity without compromising quality. From automated irrigation systems to data-driven decision-making in crop management, efficiency measures allow entrepreneurs to produce more with less, reduce waste, and maintain consistency in product offerings. By streamlining operations, agricultural entrepreneurs can meet growing demand while remaining agile and responsive to market dynamics.

Access to Markets: Expanding Reach and Distribution Channels
Scaling up requires expanding beyond local markets to reach regional, national, or even international consumers. This involves establishing robust distribution channels, forging strategic partnerships with retailers, wholesalers, and distributors, and leveraging digital platforms for e-commerce and

direct-to-consumer sales. Access to markets is not just about selling products; it's about building brand visibility, cultivating customer loyalty, and adapting marketing strategies to resonate with diverse audiences. By expanding their reach, agricultural entrepreneurs can diversify revenue streams, mitigate risks associated with localized market fluctuations, and capitalize on emerging trends in consumer preferences.

Financial Resources: Fueling Growth with Capital

Securing adequate financial resources is fundamental to scaling up in agricultural entrepreneurship. Whether through reinvested profits, loans, equity financing, or government grants, capital infusion enables entrepreneurs to fund expansion initiatives, invest in technology and innovation, and strengthen operational capabilities. Financial resources also provide a cushion against economic uncertainties and seasonal fluctuations inherent in agriculture. By accessing diverse funding sources and managing finances prudently, entrepreneurs can navigate the financial complexities of scaling up while maintaining financial sustainability and resilience.

Talent Acquisition and Team Building: Cultivating Expertise for Success

As agricultural ventures expand, so does the need for skilled talent and effective team collaboration. Scaling up involves recruiting professionals with diverse expertise, fostering a culture of innovation and continuous learning, and empowering employees to contribute meaningfully to the enterprise's growth trajectory. Building a cohesive team aligns individual strengths with collective goals, fosters a supportive work environment, and enhances organizational agility in responding to evolving market demands and industry trends. By investing in talent acquisition and team building, agricultural entrepreneurs cultivate a workforce that drives innovation, operational excellence, and sustainable growth.

Sustainability and Impact: Balancing Growth with Responsibility

Scaling up in agricultural entrepreneurship isn't just about increasing production or market share; it's about doing so responsibly and sustainably. Entrepreneurs must prioritize environmental stewardship, adopt regenerative farming practices, and uphold ethical standards throughout their supply chains. By embracing sustainability, they not only mitigate environmental impact but also appeal to consumers increasingly mindful of the origin and sustainability of their food choices. Scaling with integrity ensures long-term viability, resilience against market disruptions, and positive societal impact, positioning agricultural ventures as leaders in sustainable agriculture and responsible business practices.

Scaling up in agricultural entrepreneurship is a transformative journey that blends vision, strategic planning, operational excellence, financial acumen, talent cultivation, and commitment to sustainability. It's about turning aspirations into achievements, expanding market reach, and making a meaningful impact on global food systems. As they navigate the complexities of scaling up, entrepreneurs uphold a commitment to innovation, resilience, and positive societal change, paving the way for a sustainable and thriving agricultural future.

4. AGRIBUSINESS AND ECONOMIC DEVELOPMENT

ROLE OF AGRIBUSINESS IN SUSTAINABLE DEVELOPMENT

In the discourse of economic development, agribusiness emerges not only as a driver of growth but also as a pivotal force in fostering sustainable development. Defined as the integration of agricultural production and business practices, agribusiness encompasses a diverse range of activities from farming and livestock rearing to food processing and distribution. Its significance extends beyond economic metrics, profoundly influencing environmental stewardship, social equity, and the overall well-being of communities worldwide.

Economic Empowerment and Community Resilience

Agribusiness plays a crucial role in empowering individuals and communities economically. In rural areas, where agriculture often serves as a primary livelihood, agribusiness initiatives stimulate local economies by creating jobs and increasing income levels. For instance, consider a farming cooperative in a small village that implements modern farming techniques and gains access to markets through agribusiness partnerships. This cooperative not only improves crop yields but also provides stable incomes for its members, lifting families out of poverty and fostering community resilience.

Environmental Sustainability and Conservation

One of the most significant contributions of agribusiness to sustainable development is its impact on environmental conservation. Through sustainable agricultural practices such as agroforestry, soil conservation measures, and water management techniques, agribusiness reduces environmental degradation and promotes biodiversity. For example, a family-owned farm adopting organic farming practices not only produces healthier crops but also preserves soil health and minimizes chemical runoff into water bodies, thus contributing to long-term environmental sustainability.

Food Security and Nutrition

Agribusiness plays a critical role in ensuring food security and improving nutrition outcomes globally. By investing in research and development, agribusiness firms develop resilient crop varieties, enhance food processing technologies, and improve distribution networks. This results in increased food production and improved access to nutritious food, particularly in underserved communities. Consider a food processing company that collaborates with local farmers to source fresh produce and distribute fortified food products to schools and community centers, thereby addressing nutritional deficiencies and promoting healthier lifestyles.

Social Inclusion and Empowerment

Agribusiness initiatives also contribute to social inclusion by empowering marginalized groups such as smallholder farmers, women, and youth. Through training programs, access to markets, and financial support, agribusinesses enable these groups to participate more effectively in agricultural value chains. For instance, imagine a cooperative of women farmers who receive training in sustainable farming practices and gain access to micro-loans for purchasing seeds and equipment. This empowerment not only improves their agricultural productivity but also enhances their social status and decision-making power within their families and communities.

Agribusiness is not just about profitability; it is a catalyst for sustainable development that integrates economic growth with environmental stewardship, food security, and social equity. As you adopting innovative technologies, fostering partnerships, and promoting inclusive business models, agribusiness can significantly contribute to achieving global development goals. As we navigate the challenges of a rapidly changing world, the role of agribusiness in sustainable development becomes increasingly crucial, shaping a path towards a more prosperous and resilient future for all. Through collective efforts and shared responsibilities, agribusiness stands poised to make a profound and lasting impact on the well-being of present and future generations alike.

AGRIBUSINESS DEVELOPMENT STRATEGIES

Agribusiness stands apart as an imperative area that supports vocations as well as drives generally speaking monetary development. Characterized as the mix of horticultural creation and business exercises, agribusiness incorporates a wide range of tasks going from cultivating and domesticating animals to food handling and conveyance. The improvement of viable techniques in agribusiness is critical for amplifying efficiency, guaranteeing manageability, and cultivating comprehensive development across networks.

Upgrading Efficiency through Innovation

One of the essential techniques for propelling agribusiness improvement is cultivating development in farming practices. Embracing current advancements, for example, accuracy cultivating, aquaculture, and computerized water system frameworks can essentially improve efficiency and productivity. For example, a family-claimed ranch in a rural area takes on accurate farming strategies, using sensors and information examination to upgrade compost application and water utilization. This increases crop yields as well as diminishes input costs and limits ecological effect, showing the way that development can change conventional cultivating rehearses into high-yielding, reasonable endeavors.

Market Broadening and Worth Addition

One more basic technique in agribusiness improvement is broadening market potential, opening doors and enhancing farming items. By extending market access through organizations with retailers, exporters, and food handling organizations, agribusinesses can get steady salaries and alleviate market gambles. Consider a helping of smallholder ranchers that teams up with a nearby food handling organization to foster worth added items like sticks, sauces, and dried natural products from their excess produce. This association produces extra income streams as well as sets out business open doors and reinforces nearby economies.

Advancing Feasible Practices

Supportability lies at the core of fruitful agribusiness advancement techniques. Executing reasonable agrarian practices like natural cultivating, incorporated bother management, and protection culturing jam normal assets as well as improves item quality and market seriousness. For instance, a huge scope agribusiness partnership puts resources into sustainable power sources and embraces water-saving innovations to limit its carbon impression and conform to natural guidelines. By showing ecological stewardship, agribusinesses can assemble entrust with customers, financial backers, and administrative specialists while shielding the drawn out practicality of farming creation.

Engaging Smallholder Ranchers and Provincial Communities

Engaging smallholder ranchers and provincial networks is fundamental for comprehensive agribusiness improvement. Giving admittance to fund, horticultural preparation, and market data empowers ranchers to work on their efficiency and pay levels. Envision a grassroots association that works with rancher cooperatives, offering studios on reasonable cultivating strategies and connecting ranchers to fair-exchange markets. Through aggregate activity and information sharing, these cooperatives engage ranchers to haggle better costs for their harvests and reinforce their bartering power inside the store network.

Successful agribusiness improvement methodologies envelop a mix of development, market enhancement, manageability, and local area strengthening. As worldwide difficulties, for example, environmental change and food weakness strengthen, the job of agribusiness in practical financial advancement turns out to be progressively basic. Through essential speculations and cooperative endeavors, agribusinesses can flourish in cutthroat business sectors as well as add to building versatile and prosperous networks around the world.

PARTNERSHIP AND COLLABORATION FOR AGRIBUSINESS DEVELOPMENT

In the realm of agribusiness and economic development, the power of partnership and collaboration stands as a cornerstone for sustainable growth and innovation. The challenges faced by agricultural enterprises today are complex and multifaceted, ranging from climate change impacts to market volatility and technological advancements. To navigate these challenges successfully and to foster robust growth, partnerships among various stakeholders are not just beneficial but essential.

The Strength of Collaboration

Imagine a scenario where a small-scale farmer in a rural community dreams of expanding their operations to meet growing market demands. Alone, they face hurdles in accessing modern agricultural techniques, financial resources, and reliable markets. However, through collaboration with local agricultural cooperatives, research institutions, and government agencies, this farmer gains access to crucial resources and knowledge.

Partnerships enable the pooling of resources and expertise, thereby reducing individual risks and amplifying collective impact. For instance, agricultural cooperatives can provide economies of scale in purchasing inputs and accessing markets, while research institutions offer cutting-edge technologies and practices tailored to local conditions. Government agencies can contribute policies that support fair trade and sustainable practices, ensuring a conducive environment for growth.

Consider the success story of a cooperative of smallholder farmers in a developing country, struggling with outdated farming methods and fluctuating crop prices. Through a partnership with a nonprofit organization specializing in sustainable agriculture, these farmers received training in organic farming techniques and gained certification, opening doors to premium markets previously out of reach. This collaboration not only increased their incomes but also improved soil health and biodiversity in their region.

Furthermore, partnerships between agribusinesses and financial institutions play a crucial role in providing access to finance and insurance services tailored to the agricultural sector. In regions prone to natural disasters, such as droughts or floods, insurance partnerships can safeguard farmers' investments, offering stability and resilience against unpredictable conditions.

Beyond immediate gains, partnerships foster long-term sustainability by promoting knowledge exchange and capacity building among stakeholders. Universities and research institutions contribute by conducting studies on climate-resilient crops, sustainable farming practices, and efficient water management techniques. By disseminating these findings through partnerships with extension services and farmer associations, the impact reaches even the most remote farming communities.

Moreover, partnerships facilitate market linkages that ensure fair prices for farmers and traceability for consumers demanding transparency and ethical sourcing. Collaborations between agribusinesses and technology firms drive innovation in precision agriculture, IoT applications, and data analytics, revolutionizing farm management and productivity.

The path to achieving robust agribusiness and economic development lies in fostering meaningful partnerships and collaborations. These partnerships not only empower individuals and communities but also contribute to the broader goal of sustainable development, ensuring a prosperous future for generations to come. Together, through collaboration, we can cultivate a thriving agribusiness sector that nourishes both people and the planet.

MANAGING EXTERNAL FACTORS AFFECTING AGRIBUSINESS

In the powerful scene of agribusiness and monetary turn of events, the capacity to oversee outside factors successfully is urgent for supported development and strength. Outer factors, for example, environmental change, market instability, strategy changes, and worldwide exchange elements continually impact the agrarian area.

Understanding and proactively addressing these variables are fundamental to exploring difficulties and taking advantage of chances.

Environmental Change Versatility

Envision a cultivating local area in a district progressively inclined to unpredictable weather conditions and outrageous occasions. Ranchers here face the overwhelming undertaking of adjusting their rural practices to guarantee efficiency and supportability in the midst of changing climatic circumstances. Through interests in strong yields, water the board frameworks, and maintainable practices like agroforestry, these ranchers can alleviate gambles related with environmental change.

Organizations with environment research establishments and NGOs that work in supportable horticulture give important experiences and assets. By coordinating brilliant innovations and practices, for example, trickle water systems and dry season safe seeds, ranchers safeguard their vocations as well as add to natural preservation endeavors.

Market Unpredictability and Variation

Consider a situation where fluctuating product costs present difficulties to agribusinesses dependent on worldwide business sectors. Here, broadening turns into an essential objective. Ranchers and agribusiness ventures can moderate dangers by investigating specialty markets, esteem added items, and direct buyer commitment through rancher's business sectors or local area upheld horticulture (CSA) drives.

Also, joint efforts with monetary foundations for supporting procedures and hazard management instruments offer insurance against market vulnerabilities. These associations give admittance to custom fitted monetary items like prospects agreements or harvest protection, shielding speculations and guaranteeing monetary steadiness.

Strategy and Administrative Changes

In a world formed by developing strategies and guidelines, agribusinesses should remain educated and versatile. Changes in rural endowments, economic accords, or natural guidelines can altogether affect functional expenses and market access. Participating in backing endeavors through industry affiliations and campaigning bunches empowers partners to impact strategy choices that help reasonable horticulture and fair exchange rehearses.

Associations with government offices and lawful specialists work with consistency with administrative prerequisites while pushing for approaches that advance development and intensity. By taking part in approach exchanges and partner discussions, agribusinesses can shape great administrative conditions helpful for long haul development.

Worldwide Exchange Elements

Globalization presents two amazing open doors and difficulties for agribusinesses participating in worldwide exchange. Duties, exchange debates, and international strains can disturb supply chains and influence market access. Building vital coalitions with global wholesalers, exporters, and exchange affiliations upgrades market infiltration and diminishes reliance on unpredictable business sectors.

Moreover, putting resources into market knowledge and social comprehension assists agribusinesses with exploring different purchaser inclinations and administrative scenes in unfamiliar business sectors. Joint efforts with operations suppliers and exchange help organizations smooth out send out processes, guaranteeing opportune conveyance and consistency with worldwide guidelines.

Taking everything into account, overseeing outside factors influencing agribusiness requires proactive techniques, cooperation, and versatility. By tackling organizations with partners across areas; from ranchers and scientists to policymakers and worldwide dealers, agribusinesses can successfully moderate dangers and profit by valuable open doors in a quickly impacting world.

Through creative practices, versatile techniques, and a pledge to supportability, agribusinesses guarantee their own reasonability as well as add to financial turn of events, food security, and ecological stewardship. By embracing difficulties as any open doors for development and learning, we make ready for a versatile agribusiness area that flourishes in the midst of outside pressures, eventually helping networks around the world.

5. TECHNOLOGY AND INNOVATION IN AGRIBUSINESS

OVERVIEW OF AGTECH AND DIGITAL AGRICULTURE

Innovation isn't simply a device yet an extraordinary power reshaping the manner in which we develop, make due, and disseminate food worldwide. This upheaval, known as Agtech or Computerized Horticulture, addresses a combination of state of the art innovations and customary cultivating rehearses, pointed toward improving proficiency, manageability, and efficiency in food creation.

Embracing Advancement in Agribusiness

Envision a homestead where each part of development, from soil the board to trim observing, is brilliantly upgraded through computerized arrangements. This isn't sci-fi however a reality unfurling across ranches around the world. Computerized Agribusiness coordinates progressed sensors, satellite symbolism, and information examination to give ranchers ongoing experiences into their harvests and domesticated animals. For example, sensors implanted in the dirt can quantify dampness levels, empowering the exact water system that rations water and upgrades crop yields. Drones furnished with multispectral cameras can screen crop wellbeing and distinguish early indications of illness or supplement lacks, taking into account designated mediations that limit pesticide use and amplify reap quality.

Tending to Worldwide Challenges

In a world wrestling with environmental change and food security concerns, Agtech arises as an encouraging sign. By saddling information driven advancements, ranchers can adjust to changing ecological circumstances all the more successfully. For instance, prescient examination can gauge weather conditions, engaging ranchers to come to informed conclusions about establishing times and harvest assortments versatile to explicit environment stressors.

Besides, Advanced Farming cultivates feasible practices by upgrading asset allotment. Accuracy cultivating procedures help efficiency as well as diminish the ecological impression of horticulture. By limiting information squander and streamlining compost application, ranchers can add to soil wellbeing safeguarding and water preservation, urgent for long haul agrarian supportability.

Enabling Ranchers and Communities

Past proficiency gains, Agtech enables ranchers with information and availability.

Versatile applications convey significant experiences straightforwardly to ranchers' cell phones, crossing over data holes and democratizing admittance to agrarian accepted procedures. This computerized availability stretches out to advertise access, empowering ranchers to haggle fair costs and associate with worldwide business sectors flawlessly.

Besides, Agtech advances inclusivity by supporting smallholder ranchers in creating districts. Versatile banking and online business stages work with monetary exchanges and admittance to inputs, making everything fair and reinforcing provincial economies.

The Eventual fate of Horticulture is Digital

As we look forward, the capability of Agtech to change worldwide agribusiness is limitless. Advancements like vertical cultivating, hydroponics, and blockchain-empowered supply affixes vow to additionally reshape the scene of food creation and conveyance. By embracing these innovations, we can construct a tough food framework fit for taking care of a developing worldwide populace economically.

Agtech and Computerized Farming address a mechanical headway as well as a change in outlook by the way we feed our planet. By tackling development and embracing supportability, we can develop a future where farming flourishes as one with nature, guaranteeing food security and thriving for a long time into the future.

APPLICATION OF TECHNOLOGY IN AGRIBUSINESS MANAGEMENT

Technology has dramatically transformed the way we approach agribusiness management, bringing unprecedented levels of efficiency, precision, and accuracy to every aspect of the operation. From optimizing crop yields to streamlining logistics and distribution, technology is revolutionizing the industry and enabling farmers to achieve greater levels of success than ever before.

One of the most exciting innovations in agribusiness technology is precision agriculture. This approach involves the use of sensors, satellite imagery, and other advanced tech tools to generate highly granular data about crop growth and environmental conditions. This data can then be analyzed to make smart decisions about everything from planting times to fertilizer application rates and even water usage. By adopting precision agriculture methods, farmers can significantly increase crop yields, reduce costs, and minimize environmental impact.

Another key application of technology in agribusiness management is in logistics and distribution. With the rise of e-commerce and online marketplaces, farmers and producers must be able to quickly and efficiently deliver their products to customers across the world. Advanced transportation and logistics software can help farmers to manage everything from shipping and customs documentation to inventory management and order fulfillment. By streamlining these processes, farmers can reduce time-to-market, increase accuracy, and improve customer satisfaction.

Perhaps the most significant impact of technology on agribusiness management has been in the area of sustainability. Environmental concerns such as soil erosion, water scarcity, and pesticide pollution have traditionally been major challenges for farmers. But thanks to new technologies such as soil sensors, drones, and machine learning algorithms, farmers can now monitor their operations in real-time and identify areas where sustainability efforts can be improved. By adopting sustainable agribusiness practices, farmers can protect the environment while also reducing costs and increasing yields.

Consider a small-scale farmer who specializes in growing strawberries. By embracing precision agriculture practices, she can use soil sensors to identify the specific areas of her field that need the most water and fertilizer. She can use drones to monitor crop growth, identify disease outbreaks, and optimize her harvest schedule. And she can use advanced e-commerce software to sell her produce directly to consumers, bypassing traditional supply chains and improving her profit margins.

Technology is transforming agribusiness management in profound ways, enabling farmers to maximize efficiency, increase sustainability, and achieve higher levels of success.

AGTECH STARTUPS AND THE ENTREPRENEURSHIP ECOSYSTEM

With innovation and advancement, agribusiness tasks are quickly progressing. From crop observing to stock administration, innovation applications are changing the manner in which agribusinesses deal with their tasks. The mix of innovation has not just made agribusiness operate With innovation and advancement, agribusiness tasks are quickly progressing. From crop checking to stock administration, innovation applications are changing the manner in which agribusinesses deal with their activities. The reconciliation of innovation has made agribusiness activities more effective, however it has additionally assisted with expanding efficiency and benefit.

One illustration of innovation application in agribusiness management is accuracy farming. Accuracy farming includes the utilization of cutting edge innovation to screen harvests and soil conditions. It assists ranchers with coming to informed conclusions about their harvest management rehearses. By utilizing GPS innovation, ranchers can plan their fields precisely and dissect the information to recognize explicit regions that need more consideration. This empowers ranchers to apply manures and water just where it is required, accordingly decreasing info expenses, and water wastage, while further developing yield and harvest quality.

One more innovation application in agribusiness management is stock following. With the utilization of programming and cell phones, ranchers can effectively follow their stock, including seeds, synthetic compounds, and gear. This not just aids ranchers in dealing with their stock however it additionally empowers exact determining of data sources expenses and income, working with smoothed out tasks, and decreasing the gamble of overloading.

Innovation is additionally changing the showcasing part of agribusiness. Ranchers are currently ready to sell their items online through internet business stages. It empowers ranchers to expand their expected arrival past their customary geological market regions and offer to purchasers who are keen on their items however can't truly to come to their homestead or actual business sectors. Through web based advertising, ranchers can likewise construct their image, increment productivity, and associate with clients straightforwardly.

The utilization of innovation in agribusiness management has achieved surprising change in the business. It has driven effectiveness and efficiency as well as enabled ranchers to go with additional educated choices. The utilization of innovation has additionally helped ranchers in really overseeing stock, enhancing crop yields, and smoothing out functional methods. As innovation keeps on advancing, we can hope to see more inventive innovation applications that will upset the agribusiness industry.ons more proficient, yet it has likewise assisted with expanding efficiency and benefit.

6. CASE STUDIES IN AGRIBUSINESS MANAGEMENT AND ENTREPRENEURSHIP

SUCCESSFUL AGRIBUSINESS MANAGEMENT AND ENTREPRENEURSHIP IN DEVELOPED COUNTRIES

Agribusiness is an essential part of the economy in many created nations. Effective agribusiness the board and business venture not just add to the development of the agrarian area yet additionally support nearby networks, set out work open doors, and guarantee food security for the country. In this article, we will investigate some genuine instances of fruitful agribusiness, the board and business ventures in created nations.

One model is Denmark, a little European country with a roaring rural area. Denmark's progress in agribusiness the board and business venture can be ascribed to its attention on reasonable cultivating, areas of strength for rehearsed approaches, and backing for innovative work in the rural area. Danish agribusinesses are known for their top notch and imaginative items, for example, natural food created with negligible effect on the climate.

Another model is the US, where the farming area is a significant supporter of the economy. Effective agribusiness the board and business in the US have been driven by an emphasis on innovation and development. For instance, Accuracy Farming, an agrarian creation framework that utilizes information driven advancements to improve crop yields and decrease data sources like water and manure, has reformed the business lately. The outcome of agribusinesses like John Deere,

ADM, and Cargill can be credited to their capacity to use innovative headways for their potential benefit.

Canada is one more evolved country that has made progress in agribusiness, management and business. The nation's emphasis on an enhanced cultivating area, interest in innovative work, serious areas of strength for and strategies have added to its prosperity. Specifically, Canadian agribusinesses are known for their development of top notch and economical items like natural meat, maple syrup, and honey.

Effective agribusiness, the board and business ventures in created nations are urgent for the development of the agrarian area, and likewise, the economy. Factors, for example, economical cultivating rehearses, mechanical headways, legislative strategies and backing for research should be considered for enduring development that can reliably conquer arising issues. Fundamental to establish a climate empowers development, upholds neighborhood networks, and guarantees food security for the whole country. Danish, American, and Canadian agribusinesses give extraordinary models and motivation to other created nations hoping to develop their farming area.

BEST PRACTICES AND LESSONS LEARNED FROM DEVELOPING COUNTRIES

Fruitful agribusiness the board and business in created nations is a subject of extraordinary interest and significance as it reveals insight into the techniques that have empowered people and associations to flourish in the area. Horticulture has forever been a fundamental area in any economy, and created nations have shown that accomplishing fruitful agribusiness the board and business venture through different approaches is conceivable. This article looks at fruitful agribusiness, management and business in created nations and investigates genuine circumstances that exhibit the viability of specific methodologies.

One of the techniques that have demonstrated profoundly fruitfulness in evolved nations is development and innovation reception. These nations have outfitted innovation to further develop yields, diminish costs, further develop food handling, and make new business sectors. In the US, for example, the utilization of accurate agribusiness advances like GPS, soil dampness sensors, and robots has changed cultivating rehearses. The reception of these advances has empowered ranchers to utilize information driven dynamic cycles bringing about more significant returns, lower expenses and better benefits. Additionally, Japan has put vigorously in advanced mechanics, computerization, and man-made brainpower in agribusiness. These advances have empowered the country to manage its property shortage challenges while keeping up with high efficiency levels.

One more technique that has worked in evolved nations is the execution of economical agribusiness rehearses. The world is turning out to be all the more earth cognizant, and there is a developing interest for maintainability in farming. Effective agribusinesses in nations like Denmark, Sweden, and the Netherlands have carried out maintainable cultivating rehearsals. For instance, the Dutch have embraced feasible nursery cultivation, and they have turned into the world's second-biggest exporter of vegetables. The country's nursery ranchers utilize inventive methods like intensity stockpiling and lighting to increment yields while diminishing energy utilization. They additionally utilize regular vermin control strategies and coordinated water system frameworks to ration water.

Also, fruitful agribusiness the board in created nations has been portrayed by esteem expansion and broadening. Esteem option includes overhauling farming items and handling them into higher worth added items. This methodology has been embraced in New Zealand where the public authority has given motivations to ranchers to wander into esteem added items. For instance, the country's dairy ranchers have enhanced cheddar, margarine, and milk powder creation, which has brought about essentially higher benefits. Moreover, there is likewise a developing pattern of agro-the travel industry in created nations, where ranchers have differentiated their revenue streams by giving vacationer encounters by means of homestead visits, convenience, and friendliness.

Effective agribusiness management and business in created nations require a blend of development, supportability, and expansion. The genuine circumstances featured in this article exhibit the viability of these systems in further developing efficiency, lessening costs, and expanding benefits. The reception of these systems in less evolved nations could prompt comparable advantages in the farming area, which is fundamental in making an economical future for the planet

7. CONCLUSION AND FUTURE TRENDS IN AGRIBUSINESS MANAGEMENT AND ENTREPRENEURSHIP

SUMMARY OF KEY LEARNING POINTS

As we arrive at the finish of our conversation on what's to come, patterns in agribusiness, management and entrepreneurship, it is fundamental that we ponder the key gaining focus from our past conversations. These are fundamental important points that can assist us with exploring the difficulties and chances of the quick changing agribusiness scene.

One of the fundamental learnings is the significance of innovation in agribusiness to management. The reception of accuracy in horticulture and related innovations can further develop crop yield, lessen costs, and work on the viability of supportability drives. For example, accurate horticulture innovations can assist ranchers with gathering information about soil ripeness, weather conditions, and yield development, permitting them to enhance the use of manures, herbicides, and pesticides. By embracing innovation like this, ranchers increment yields and get a good deal on inputs.

One more basic learning point is perceiving the significance of manageability in agribusiness the board. With the impacts of environmental change turning out to be more obvious step by step, there's an incredible need to investigate feasible agribusiness rehearses further. Economical agribusiness practices like yield revolution, utilizing locally accessible assets, and water reaping are fundamental in ensuring that we add to a better climate and guarantee food security for individuals later on ages.

Entrepreneurship venture and development inside the agribusiness area have likewise been recognized as a basic component. It is fundamental to put resources into imaginative plans of action that can possibly offer new and intelligent fixes to the current difficulties of the horticultural area. By making creative arrangements, for example, further developed coordinated factors and a web based business stage, business people can assist ranchers with defeating the basic issues resolving the contemporary rural issues.

In a genuine application, the manageability advocate, Jane is an agribusiness business person that utilizes agroforestry rehearses in her espresso ranch. Through establishing trees and through soil air circulation, and fertilizer application, Jane has had the option to further develop soil wellbeing and surface, shielding her yields from bugs and illnesses and expanding her efficiency. Jane, through her inventive practices, guarantees a solid climate and adds to the nearby economy by utilizing countless specialists.

At long last, the agribusiness scene is as yet confronting vulnerability. Accordingly it is fundamental to be continually mindful and versatile of changes in the business. Through systems administration with specialists, ranchers can stay up with the latest with new turns of events. Integrating innovative work projects to manage the difficulties the climate presents can be a helpful method for remaining on the ball.

As we see future patterns in agribusiness, management and business venture it is fundamental to recall that innovation, manageability, and business venture are essential in getting a handle on the future open doors and tending to the difficulties inside the farming area. By gaining from foundations, embracing development, and maintainability techniques as business person and ranchers, the fate of agribusiness the board is splendid with developments towards a greener economy.

TRENDS AND INNOVATIONS IN AGRIBUSINESS

As of late, the agribusiness business has seen a progression of creative patterns and upgrades, making the area more proficient, useful, and cutthroat. These patterns and advancements are rethinking the eventual fate of agribusiness and how we produce food in our quickly developing world.

Perhaps the main pattern in agribusiness is economical horticulture. As the world plans to move towards reasonable practices, the horticulture area isn't to be abandoned. With the rising attention with the impacts of environmental change, purchasers are requesting harmless ecosystem items made through practical cycles. Natural cultivating, crop revolution, and protection culturing rehearses are only a couple of the feasible cultivating strategies utilized by ranchers. This pattern is supposed to develop as an ever increasing number of individuals focus on a reasonable way of life.

One more pattern in agribusiness is the advancement of specialty crops. Specialty crops like colorful natural products or vegetables have become progressively famous lately, offering preferable benefits over conventional harvests. With the ascent in populace, there is likewise an expansion popular for food, particularly good food choices, making specialty crops a rewarding and open door for ranchers to satisfy this need. A portion of these harvests incorporate blueberries, quinoa, and kale.

Accuracy agribusiness is another significant development that has upset the agrarian scene lately. Accuracy horticulture utilizes innovation to streamline ranch tasks, permitting ranchers to plant precisely, oversee information sources, and screen crops from a distance. This incorporation of innovation has essentially further developed yields, made the business more proficient, and possibly decreased input costs. The utilization of robots, GPS programming frameworks, soil sensors and other computerized devices for cutting edge investigation are providing ranchers with a superior comprehension of the dirt, crop condition, supplements and the general homestead execution.

Vertical cultivating is another development rapidly disturbing the conventional cultivating rehearses. Vertical cultivating involves developing harvests in stacked layers in completely controlled conditions, making them resistant to outside climate varieties. This approach resolves the issue of food security and shortage. With the expansion in globalization and urbanization, the requirement for privately obtained food items turns out to be progressively significant as transport turns out to be more troublesome because of environmental change, normal fiascoes and financial difficulties.

As the world keeps on moving towards an all the more mechanically progressed future, the horticultural area is likewise growing with recent fads and developments. Manageable cultivating rehearses give a greener and more eco-accommodating way to deal with crop creation. Speciality crops offer more significant yields for ranchers by taking care of the rising interest for good food choices. Accuracy horticulture and vertical cultivating are now yielding great outcomes by expanding efficiency and expanding the stock of food items. What's in store looks encouraging for the horticulture area, as it proceeds to advance and adjust to the changing requests of the world.

RECOMMENDATIONS FOR POLICY MAKERS AND STAKEHOLDERS.

As we close our conversation on future patterns in agribusiness, management and entrepreneurship, featuring a few vital proposals for strategy creators and stakeholders is fundamental. These proposals will help in establishing a favorable climate for development in the horticultural area, support little and medium-sized organizations, and guarantee food security for our nations.

There, right off the bat, is a requirement for strategies that help maintainable cultivating rehearsals. Many created nations have been executing approaches that are equipped towards natural preservation and the decrease of carbon impressions. Strategy producers ought to expect to help limited scope ranchers with these approaches by giving motivators and endowments that advance the reception of manageable practices like harvest expansion, crop pivot, and low-till cultivating.

Furthermore, there is a requirement for strategies that support innovative work in the rural area. The rural business is quickly developing, and there is a requirement for persistent development towards supportable practices for cultivating and better food creation. Partners ought to put resources into innovative work of trend setting innovations that empower more proficient food creation, and strategy producers ought to energize continuous advancement in the area by offering financing valuable open doors.

Thirdly, putting resources into the schooling and preparing of ranchers and partners in the rural sector is fundamental. This venture will assist with achieving better quality items, supportable practices, and more proficient food creation. There is a requirement for strategies that give preparing and schooling potential chances to limited scope ranchers, so they can stay aware of the most recent patterns and innovations to guarantee their prosperity.

Fourthly, approach creators ought to focus on interest in frameworks like vehicle, storerooms, and correspondence organizations to work with food dispersion and showcasing. This will assist with defeating the difficulties of distance and planned operations that hamper the development of the horticultural area. By putting resources into foundation, ranchers can essentially lessen their creation costs, and successfully arrive at new business sectors with their items.

Finally, notwithstanding crises, approach creators ought to go to proactive lengths to help neighborhood makers and independent ventures. Unique approaches pointed toward balancing out the market and decreasing the effect of calamities ought to be underway.

The rural business is an essential piece of the economy, with huge commitments to food security, pay age, and open doors. To accomplish feasible development and improvement in the area, strategy creators should be prepared to offer monetary help and advance arrangements that support economical cultivating practices, innovative work, schooling and preparing, foundation advancement as well as catastrophe emergency courses of action. Partners like financial backers, ranchers, and agribusinesses should likewise assume a part in supporting the execution of these strategies, as well as leading developments towards reasonable practices. Working on the whole and putting resources into meditations for a superior farming industry will hence reinforce the economy.

www.ingramcontent.com/pod-product-compliance
Lightning Source LLC
Chambersburg PA
CBHW071933210526
45479CB00002B/657